WHITE LAKE TOWNSHIP LIBRARY
7527 E. HIGHLAND RD.
WHITE LAKE, MI 48383
(248) 698-4942
#73
SEP 27 2018

EVOLVING TECHNOLOGY

THE EVOLUTION OF TRANSPORTATION TECHNOLOGY

J. LAKE

Published in 2019 by Britannica Educational Publishing (a trademark of Encyclopædia Britannica, Inc.) in association with The Rosen Publishing Group, Inc.
29 East 21st Street, New York, NY 10010

Copyright © 2019 The Rosen Publishing Group, Inc. and Encyclopædia Britannica, Inc. Encyclopædia Britannica, Britannica, and the Thistle logo are registered trademarks of Encyclopædia Britannica, Inc. All rights reserved.

Distributed exclusively by Rosen Publishing.
To see additional Britannica Educational Publishing titles, go to rosenpublishing.com.

First Edition

Britannica Educational Publishing
J.E. Luebering: Executive Director, Core Editorial
Andrea R. Field: Managing Editor, Compton's by Britannica

Rosen Publishing
Bailey Maxim: Editor
Nelson Sá: Art Director
Brian Garvey: Series Designer
Tahara Anderson: Book Layout
Cindy Reiman: Photography Manager
Ellina Litmanovich: Photo Researcher

Library of Congress Cataloging-in-Publication Data

Names: Lake, J., author.
Title: The evolution of transportation technology / J. Lake.
Description: New York : Britannica Educational Publishing, in Association with Rosen Educational Services, 2019. | Series: Evolving technology | Includes bibliographical references and index. | Audience: Grades 5–8.
Identifiers: LCCN 2017044411| ISBN 9781538302873 (library bound : alk. paper)
| ISBN 9781538302880 (pbk. : alk. paper)
Subjects: LCSH: Transportation—History—Juvenile literature. | Transportation engineering—History—Juvenile literature.
Classification: LCC TA1149 .L345 2019 | DDC 629.0409--dc23
LC record available at https://lccn.loc.gov/2017044411

Manufactured in the United States of America

Photo credits: Cover (left to right) Heritage Images/Hulton Archives/Getty Images, Martin Lehmann/Shutterstock.com, Denis Belitsky/Shutterstock.com; cover, back cover, pp. 4-5 (background), 6, 19, 30, 44 spainter_vfx/Shutterstock.com; p. 5 © Hungarian National Museum, Budapest; photograph, Kardos Judit; p. 7 Bildagentur-online/Universal Images Group/Getty Images; pp. 10, 21 © Encyclopædia Britannica, Inc.; p. 12 © Edward S. Curtis Collection/Library of Congress, Washington, D.C. (cph 3a16196); p. 14 Buyenlarge/Archive Photos/Getty Images; p. 16 Photos.com/Thinkstock; p. 23 © Photos.com/Jupiterimages; pp. 26-27 Photo 12/Alamy Stock Photo; p. 28 Universal History Archive/Universal Images Group/Getty Images; p. 32 Stefano Bianchetti/Corbis Historical/Getty Images; p. 35 © Library of Congress, Washington, D.C. (digital file no. 00658u); p. 36 New York Daily News/Getty Images; p. 40 © U.S. Navy photo; p. 42 © MSFC/NASA; p. 45 Peter Titmuss/Shutterstock.com; p. 47 Grzegorz Czapski/Shutterstock.com; p. 49 Fedor Selivanov/Shutterstock.com; pp. 50–51 NASA/Goddard Space Flight Center; p. 55 Lee Snider Photo Images/Shutterstock.com.

CONTENTS

Introduction 4

CHAPTER 1
Early Transportation 6

CHAPTER 2
Steam Power, Railroads,
    and the Bicycle................................... 19

CHAPTER 3
Transportation Revolutions 30

CHAPTER 4
The Future of Transit 44

Glossary 58
For Further Reading 61
Index... 63

INTRODUCTION

From camels and rafts to rockets and submarines, transportation technology has grown and developed alongside human civilization. At its most basic level, the purpose of transportation is to move people and goods more easily or across greater distances than would be possible on foot. But through the ages, vital communications have also gone hand-in-hand with advances in transportation, including the royal messengers running thousands of miles along the Inca foot highway five hundred years ago, as well as today's jet-setting elites and governmental envoys, carrying briefcases with billion-dollar trade secrets and negotiations to end armed conflicts.

From ancient times transportation technology changed slowly or in spurts. By the 1800s, however, steam-powered machines helped people span distances like never before. More efficient diesel and gasoline-powered engines dominated the 1900s, followed by electric and computerized transportation technologies in our own time.

Not all people have equal access to transportation technologies, however. Wealthier citizens can afford the most up-to-date private cars, although major cities are in desperate need of affordable and responsible public transit for people of all income

A mug-sized model of an early wheeled cart. The use of wheels helped carry people, crops, and supplies farther and with less effort than ever before, aiding in the evolution of farms and towns.

levels, as well as for people with disabilities. Moving further into the twenty-first century, transportation technology must balance speed and efficiency with more holistic concerns, such as environmental sustainability and more equal distribution and access.

Chapter 1

EARLY TRANSPORTATION

Early humans relied on walking to get from place to place. With the advent of agriculture, domesticated animals provided a source of food and clothing. Eventually people saw greater potential in some animal species, which they harnessed to provide faster and easier travel.

In the Middle East, camels have been used as transportation for thousands of years. They are often called "the ships of the desert" because they provide such an ideal form of transportation, as they are able to go without water for days and can carry people or freight. Other animals were domesticated for transportation in different regions.

Early peoples also found that putting heavy loads on sledges enabled them to slide heavier objects along the ground with less effort than needed to carry

Camels have been used for thousands of years in Egypt and other parts of the Middle East to carry people and goods. The animals are well suited for their desert environment.

The Evolution of Transportation Technology

them. Ultimately, they put the sledges on tree limbs as rollers, which led to the development of wheels on an axle.

USING ANIMALS FOR TRANSIT

In ancient times people began experimenting with domesticated animals to help carry passengers and goods. Dogs were domesticated at least fifteen thousand years ago, being used both as guards and as a food source. At some point people began to use teams of dogs to carry loads, a tradition that is still seen in dogsledding today.

Camels were used in ancient Africa and Asia to transport people and their belongings. Domesticated horses were used to pull chariots long before they were saddled for individual riders. Horses became common in Egypt around 1700 BCE, when they were brought over from Syria. Horses could travel faster than camels but were also more prone to exhaustion in the desert heat.

Relatives of the camel, the llama and alpaca, have been used in South America for thousands of

Early Transportation

years to help transport goods. Meanwhile, elephants were used in Asia and Africa for transport, for hauling logs, and for warfare. The larger African elephants were more temperamental and harder to control. In the third century BCE the Carthaginian general Hannibal invaded Italy by crossing the Alps with an army that included thirty-seven elephants. After several battles in difficult terrain, however, only his elephant survived.

INVENTING THE WHEEL

Logs were used as rollers long before the invention of the wheel. The first wheels, then, were likely cut from logs. It is not known exactly when the wheel first came into existence, but a Sumerian artifact from around 3500 BCE shows a sledge with wheels, and the Mesopotamians were using the wheel to spin pottery about the same time. As the technology improved over time, wheels were attached to a fixed axle and could make easier turns.

Around 2000 BCE, Egyptian chariots featured a spoked wheel that was more sophisticated than a

Improvements on the wheel

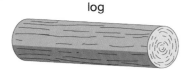
log

The earliest wheels were disks cut from a log.

cleat
peg
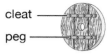

Wheels of larger diameter were made from three boards, held together with cleats and wooden pegs.

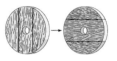

Stronger wheels were made by sandwiching two disks, with the wood grain crossing at right angles.

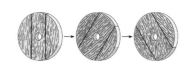

Further improvement was made by sandwiching three disks, with the wood grain crossing at angles of 60 degrees.

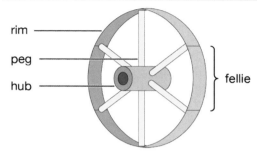
rim
peg
hub
fellie

The middle boards of the disks above led to the development of six-spoke wheels with rims and large hubs to keep the wheel from wobbling

© 2012 Encyclopædia Britannica, Inc.

> Wheels developed over time. They began as rough pieces cut from logs and were made stronger and easier to operate with the use of layered wooden disks, followed by spokes (pegs) and rims.

simple wheel cut from a log. Spoked wheels were stronger and wobbled less, but they were suitable only for lighter loads.

The Roman Empire, which covered large parts of Europe, northern Africa, and the Mediterranean islands, needed advanced transportation methods for far-flung trade and communications. The Romans thus developed vehicles for a variety of purposes, from covered carriages to farm carts, as well as lightweight racing chariots for entertainment. They built a vast paved highway system outward from Rome around two thousand years ago.

TECHNOLOGY OF INDIGENOUS PEOPLES

Although there is no record of the wheel in North or South America before the arrival of Europeans, the Incas built a road system through the central Andes Mountains that stretched as far as 2,250 miles (3,624 kilometers). The thousands of stone steps on this Inca highway were suitable for human messengers and porters, as well as for llamas carrying packs weighing up to about 70 pounds (32 kilograms).

In North America, subarctic tribes made wooden sleds called toboggans to move heavier articles. Snowshoes also improved conditions for travelers walking long distances in harsh winter weather. American Plains Indians designed a travois, a type of sled pulled by people or by sled dogs. After the arrival of Europeans, Native Americans also began to ride horses and to use them to pull travois.

The Evolution of Transportation Technology

BOAT TRAVEL BEGINS

To make boats, native peoples of the Americas used materials at hand, including dugout tree trunks and animal hides. Inuit made animal-skinned kayaks to

Native peoples of the Americas, such as the Inuit, fashioned early boats out of animal skins and tree logs that allowed them to fish farther away from shore.

Early Transportation

fish, while American Plains Indians used bullboats, wrapping an animal hide over a wooden frame.

In the Pacific Ocean, the Polynesians proved themselves the early master navigators of the world. They explored thousands of miles of ocean territory and settled distant islands, from New Zealand to Hawaii and Easter Island, using canoes which they enlarged and equipped with sails. They navigated using the stars, knowledge of ocean currents, and maps they made from reeds and shells.

Early Egyptians did not have large trees for dugout canoes; instead, they constructed cargo ships using smaller pieces of wood tied together, which could carry up to 350 tons (317 tonnes) of stone. The Phoenicians put together planked wooden ships, using Lebanon cedar to create vessels with an elaborate ribbed structure. Using oars and sail-power on their fleets, the Phoenicians controlled trade in the Mediterranean for centuries.

The Greeks invented the cross-staff and the astrolabe, which helped determine location at sea. By 1000 CE, Viking ships took on a distinctive form, with pointed bows and overlapping oak planking called clinker building. Vikings also used oarsmen to power their ships.

The Evolution of Transportation Technology

The magnetic compass, developed in the 1100s in China and Europe, gave navigators a more accurate way of determining the direction of travel. Soon after, Italians began making charts called *portolanos* to map the coastlines. By 1450, Europeans had the means to build large ships with up to three masts that could travel long distances. These boats, com-

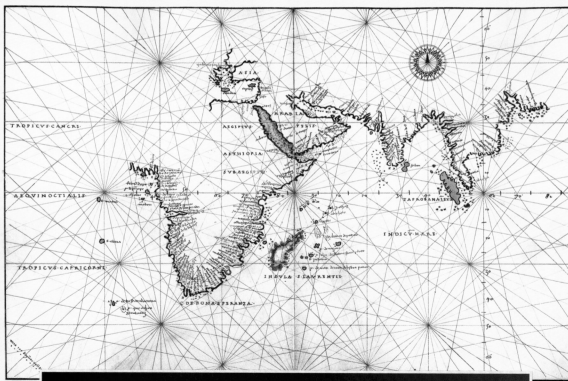

In the fourteenth century, Italians started making navigational charts, or *portolanos*, to map coastlines. These charts helped navigators avoid hazardous rocks and find safe harbors.

bined with the compass, allowed voyages of exploration around the world.

THE AGE OF EXPLORATION

The European Age of Exploration began in the 1400s when explorers were sent to find new trade routes to get spices and silks from Asia. Blocked from traveling through the eastern Mediterranean and other parts of Europe, Portugal and Spain began to seek out oceanic routes southward and westward. They explored the coast of Africa, attempting to reach the East Indies via the Indian Ocean. Christopher Columbus then used sailing ships called caravels to cross the Atlantic Ocean. This led Spanish conquistadores to explore and conquer the Americas, causing the end of the Aztec and Inca empires.

These voyages were the beginnings of global exchange, as Europeans brought technologies such as steel weapons, guns, and ships to the Americas and brought back maize, cacao (which is used to make chocolate), avocados, tomatoes, potatoes, tobacco, and gold from the Americas. With more nutrient-rich food, Europeans could boost their populations, which in turn gave them more power to

The Evolution of Transportation Technology

THE COST OF SHIPPING: FORESTS

During the Age of Exploration, Europeans built increasingly massive wooden sailing ships. A tremendous amount of wood was needed for ship frames and planks, with builders favoring thick, old tree trunks for the hulls. As a result, oak forests of Great Britain and northern Europe were chopped down, and large areas were deforested. As more and more ships were constructed, builders started

During the Age of Exploration, large ships were built out of oak. These ships, with up to three masts, could carry people, provisions, and merchandise across vast oceans.

Early Transportation

looking to Asia and the Americas. They used the oak of North America and teak wood from India to keep constructing ships. The white pines of New England and the Douglas fir trees of the Pacific Northwest were especially valued for ship masts. By the 1850s, Europeans started switching over to iron as a major shipbuilding material, but wood was still needed. In order to heat iron to shape it, firewood and, later, coal were used to create high temperatures.

dominate other groups. At the same time, the Europeans unwittingly brought diseases that caused millions of people to sicken and die in the Americas.

TRANSPORTATION AND HUMANITARIAN CRISES

European voyagers often brought calamity to the cultures they visited. European sailors, fortune-seekers, and settlers looted the land and murdered, kidnapped, and brought sickness to the indigenous populations. Portuguese explorers in the 1500s captured increasing numbers of Africans and brought

The Evolution of Transportation Technology

them back to Europe as slaves. As Europeans established outposts in the Americas, they enslaved many of the indigenous inhabitants. Even those who were not enslaved were often abused. Later, Europeans started bringing shiploads of African slaves to work the plantations and mines. Europeans were able to consolidate wealth by creating these slave-based plantation systems, where slaves provided labor to produce raw materials, such as sugarcane and cotton, that were sent back to Europe.

Meanwhile, large numbers of European farmers and laborers were enticed to the "New World" as indentured servants, who signed away years of their lives in the hope of providing a more secure future for their families. Tragedies such as the Irish potato famines and wars in mainland Europe pushed many others to emigrate as well. Diseases were rife on these ships as sanitation technologies lagged behind. Many sailors and passengers suffered from scurvy, when long voyages caused a lack of fresh vegetables and fruits in their diets.

Chapter 2

STEAM POWER, RAILROADS, AND THE BICYCLE

The Industrial Revolution, which began in the late 1700s and gained strength in the 1800s, set off a series of transformations that changed the way people lived, traveled, and worked. Factories in England started producing textiles, which caused machines to replace the work of artisans. However, working conditions in these textile factories, which often employed women and children, were very harsh. In England, people had started moving to the cities because wealthy British landowners were taking land away from peasants. At the same time, advances in food production meant that cities could support denser populations.

Yet not every region immediately followed this industrial path. By the early 1800s, Asian countries had not started to industrialize, although Asian cities contained about two-thirds of the world's urban population, led by Edo (now called Tokyo) with over one million residents. Other regions, particularly those colonized by European powers such as Africa and South America, did not have the access to investment funds needed to change their infrastructure.

STEAM POWER

The first factories of the Industrial Revolution were water-powered, which meant their waterwheels had to be located on or near flowing streams. Yet coal production and advances in steam technology brought factories to far-flung locations. Steam engines created power by heating water until it became steam and then using the high pressure of the steam to push pistons into motion. As more compact steam engines were invented, they were used to drive ships and then railroad engines.

Sailing ships could move thousands of people in flotillas and convoys, but they were often slow and unreliable, especially during rough winds or storms at sea. Meanwhile, on land, wagons and carts pulled by donkeys, oxen, or horses had been used for centuries to move people from place to

Steam Power, Railroads, and the Bicycle

place. But with the arrival of the steam engine, new options came about for people to travel faster and more comfortably.

In 1807, the American inventor Stephen Fulton's steamer, *North River Steamboat*, made the first successful trip from New York City's harbor to Albany and back, in spite of skeptics who did not think the trip was possible. Steamship technology spread to England but was probably most influential

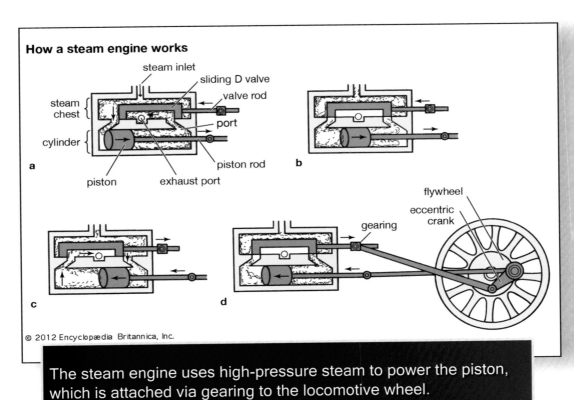

The steam engine uses high-pressure steam to power the piston, which is attached via gearing to the locomotive wheel.

21

The Evolution of Transportation Technology

in the United States on the Mississippi River, where steam-powered paddleboats helped people settle along the riverbanks, all the way from Minnesota to New Orleans.

RAILROADS SPRING UP

Early concepts for railroads were used from the mid-1500s, including horse-drawn cars on tracks. Horse-drawn trams also helped many cities and their suburbs grow in the 1800s. But it wasn't until steam engines were improved that modern railroads began to spread. Like the steamship, railroads helped move people and goods within countries, although the railroad also linked European countries together. Modern railroads were first established in Great Britain, a densely populated country with ample coal and financial resources. Richard Trevithick came up with the first successful steam locomotive in Wales in 1804. The Stockton and Darlington railroad, established in England in 1825, was the first steam-powered railway open to the general public.

Railroads in the United States faced significant challenges, as the country was much bigger, with a sparse central and western population and, initially,

Steam Power, Railroads, and the Bicycle

This locomotive, the *Rocket*, was the fastest in 1829. Railways helped link long stretches of countryside and allowed people and supplies to travel faster on land.

limited investment capital. The Baltimore and Ohio Railroads and the South Carolina Railroad both began in 1830, carrying supplies to ports. During the US Civil War (1861–1865), north-south railroad routes were constructed to serve the advancing northern armies.

The Evolution of Transportation Technology

After the war, the Union Pacific and Central Pacific railroads were joined. This formed a truly east-west transcontinental railroad, making travel across the entire United States much faster and easier. These advances came at significant human cost, however, as work on the transcontinental railroad was difficult and dangerous. Chinese immigrant laborers contributed immensely to the building of the railroad, with about fifteen thousand workers employed, in spite of rising anti-immigration fervor amongst other Americans.

CITY TRANSIT

As steamboats and railroads changed the shape of long-distance travel and trade, city transportation systems were changing everyday travel. In 1832 in New York City, trams were pulled by horses on tracks, but by the 1870s, New York's trams were powered by steam. The first underground subway was built in London in 1863, using steam locomotives despite the heavy black smoke they produced. Moving large groups of people quickly from place to place allowed for people to take jobs outside of their immediate neighborhoods, although laborers still generally lived in factory towns. Public transit also brought people together in new and unexpected ways, as city-dwellers from many paths of life used trams and subways.

THE BEGINNINGS OF THE BICYCLE

An early form of the bicycle was invented by Baron Karl von Drais de Sauerbrun at the beginning of the 1800s. The *draisienne*, named after its inventor, was a two-wheeled vehicle that the rider pushed along while walking or running. The prototype of the bicycle corresponded with growing European middle classes who had more time for recreation.

In the middle of the nineteenth century, bicycle technology advanced in Europe, although bikes were known at that time as *velocipedes*. In 1840, a Scottish blacksmith, Kirkpatrick Macmillan, built a simple velocipede with pedals that could remain upright, unlike previous models. The velocipede craze took off in the 1860s, after pedals were connected to the machine's front wheel. Mechanic Pierre Lallement exhibited the updated machine in Paris, France. The Olivier brothers popularized the velocipede throughout the entire country in 1865, when they traveled

The Evolution of Transportation Technology

Johnson, the First Rider on the Pedest...

The earliest forms of the bicycle did not have pedals. In the mid-nineteenth century, a predecessor of the modern bicycle started to become popular. It used two wheels and pedals.

Steam Power, Railroads, and the Bicycle

over 500 miles (804 km) from Paris to Marseille.

The term *bicycle* started to replace *velocipede*. Bicycle popularity spread to the United States in 1868 but then fell once people realized bikes could not travel long distances because of roads plagued by stones, potholes, and muddy ruts. Different bike models continued to be developed, in any event. Among them was the early motorbike, invented when an Englishman named Edward Butler added a motor to a tricycle. The growing popularity of bicycling led to the pavement of roads in the 1890s in Europe and the United States.

Another interesting side effect of bicycles was its effect on women's clothing. Women who wanted to bike began to wear bloomers, or loose pants that tightened at the ankle. This was considered scandalous by some people. It

The Evolution of Transportation Technology

was certainly a sharp departure from the restrictive clothes typical of the era.

RAILWAYS AND STEAMSHIPS GO GLOBAL

By the end of the 1800s, railroad technology was spreading across the world to aid plantation production, mining, and industrialization. In the Belgian Congo, railroads were built to transport rubber and other resources from the interior of the country to

A railway in Rio de Janeiro, Brazil. Railroad technology spread across the world in the late nineteenth century to serve urban commuters and to move raw materials to industrial centers.

its ports. In South America and in Mexico, railroads were also built to speed up the export of crops and raw materials.

In 1819, the first transatlantic crossing of a steam ship, the *Savannah*, was successful. Yet it had to use sail power to supplement steam power. Still, the Industrial Revolution continued to increase the pace of life and production in many regions. Other ships began regular transatlantic routes in 1838, and in 1845 the first iron-hulled steamer with a screw propeller crossed the Atlantic. Steamships were not considered completely reliable until the 1880s, however, when a surge of European immigrants reached the Americas. Transportation was poised to go through dramatic changes in the 1900s.

Chapter 3

TRANSPORTATION REVOLUTIONS

Societies changed rapidly in the twentieth century, leading to transportation revolutions. In no other time period have so many crucial transit changes occurred. As new factories and cities continued to grow, causing rapid shifts in production patterns, new transportation options were needed to move people and goods more efficiently and with greater comfort. The 1900s saw the spread of the car and the airplane, as well as even more advanced vehicles such as spaceships, taking people farther than ever before.

Industrialized nations had funds to invest in newer technologies. Developing nations did not have access to many of the same advances, unless they paid a far higher premium to bring in European or US engineers and equipment. Even within industrialized countries, access to technology was not equitable. For instance, lower-income people in the urban areas of the United States and Europe had to rely on sometimes-inefficient bus systems, while middle- and upper-class citizens had greater freedom because they could afford cars to get them from place to place.

Warfare created more inequities, both from devastation to certain factories and cities and from rapid advances in transportation. Armed forces depended on rapid and long-distance transit, both for their combat operations and for simple supply lines. In particular, World War I (1914–1918) and World War II (1939–1945) saw widespread death and destruction, but they also brought amazing advances in airplanes, cargo ships, trucks and other automobiles, and even more advanced transportation technologies such as jets and rockets.

THE AUTOMOBILE HITS THE STREETS

The modern car originated from steam-powered cars, which about one hundred small companies produced in the 1890s and 1900s. The American brothers Francis and Freelan O. Stanley built the most

The Evolution of Transportation Technology

famous model, named the Stanley Steamer. Inventors also experimented with alternatives, and about thirty-five thousand electric cars were sold though they could not travel long distances.

A practical gas engine was designed and built by Étienne Lenoir of France in 1860. Newer models were created by Austrian and German inventors in

Steam-powered cars were early predecessors to the modern car. They lacked widespread popularity because they were loud and heavy, and an inexpensive version was not offered to the public.

the 1870s and 1880s. In the 1890s, J. Frank and Charles E. Duryea sold the first successful gasoline-powered automobiles in the United States. In 1896, Henry Ford drove his first car in Detroit. Meanwhile, Ransom E. Olds started mass-producing his Oldsmobile cars, producing a batch of four hundred cars that sold for $650 each. Ford came out with his Model T car in 1908, which sold more than fifteen million units in the next twenty years.

THE WRIGHT BROTHERS TAKE FLIGHT

The 1900s saw airplanes evolve from light wood-and-cloth craft to sizeable rotor-engine airplanes, followed by the massive passenger and cargo jets we know today. Although early twentieth-century inventors were the first to design airplanes that could actually fly, the idea was not new. Leonardo da Vinci had studied the flight of birds and made sketches for planes and helicopters in his notebooks. In subsequent centuries, other inventors experimented with hot air balloons, gliders, and contraptions with bird-like wings.

At the time the automobile industry was just beginning, Orville and Wilbur Wright flew the first

The Evolution of Transportation Technology

successful engine-powered plane in 1903 near Kitty Hawk, North Carolina. They had a stronger connection with bicycles than automobiles, however, since the Wright Brothers earned their money building and selling bicycles in a shop in Ohio.

After the flight near Kitty Hawk, planes were developed at a rapid pace, especially in the United States and Europe. During World War I, planes were used for observation, dueling, and bombing. Engineers also began attaching machine guns to military planes. Meanwhile, continuous delivery of airmail started within the United States. Yet despite their light frames and their ability to glide to a landing on grassy fields, these early airplanes were notoriously dangerous. Deaths were common among early pilots, partly owing to poor weather forecasting and the danger of storms to small aircraft.

After World War I ended, there wasn't as much demand for trained pilots outside of some armed forces. But civilian aviators were eager to fly farther than they had before. In 1927, Charles Lindbergh made the first flight across the Atlantic Ocean in his plane, *Spirit of St. Louis*, which was heralded around the world as a major achievement. Five years later, Amelia Earhart became the first woman to make the

A photo of the Wright brothers flying over a field. The brothers were the first to carry out a successful engine-driven airplane flight, on December 17, 1903.

solo Atlantic voyage, inspiring generations of adventurous young people.

Airplanes briefly met with competition from airships, including large dirigibles and zeppelins, which were used on bombing raids during World War I. In the 1920s and 1930s, airships carried hundreds of passengers across the Atlantic Ocean. Yet some designs were filled with highly flammable hydrogen gas. A series of fires and crashes, including the

Charles Lindbergh standing in front of his plane, the *Spirit of St. Louis*. Lindbergh completed the first transatlantic airplane flight, from Long Island, New York, to Paris, France, in 1927.

shocking *Hindenburg* disaster in 1937, made airships commercially obsolete even as airplane designs were rapidly advancing.

LAND, AIR, AND SEA TRAVEL EXPANDS

In the 1920s the automobile started becoming more commonplace in cities and suburbs. To serve the

TRANSPORTATION IN WORLD WAR II

World War II spurred countries to advance their military technology, which caused advances in transportation. German bombers attacked London in 1940, and Allied bombers and their fighter escorts quickly attempted to even the score. Eventually, flights of heavy bombers obliterated entire cities in Germany and Japan.

Later in the war American, Japanese, and German manufacturers began racing to create powerful planes with jet engines. Similar jets were later used in commercial planes, allowing air travel to become faster and more affordable for the citizens of various countries.

Developments in plane engines also carried over to the engineering of cars. Car engines became more robust, and functions such as power brakes and better steering were developed. The US Jeep military vehicle, which took its name from the phrase "g.p.," or "general purpose," was used in World War II to carry soldiers as well as anti-tank weapons and machine guns. Later, the Jeep became a popular car for civilians.

(Continued on the next page)

The Evolution of Transportation Technology

(Continued from the previous page)

The modern tank, which had appeared in simpler forms during World War I, was used extensively on World War II battlefields. Tanks were outfitted with wireless radio technology, that is, with either FM or AM radio receivers and transmitters, so they could be commanded as part of a mobile fighting unit. Many of these wartime inventions were later used in civilian engines, powering farm tractors and diesel trucks.

expanding middle-class population, the Federal Aid Road Act provided funding for a US highway system. In Europe, express highways such as the German *autobahn* and the Italian *autostrada* were built in the 1920s and 1930s. Car travel changed how people arranged their lives because a car-owning family could live farther away from work and urban centers. Once highway systems were in place, trucking became a vital way for goods to move across longer distances.

After World War II, commercial air service became affordable and covered longer distances owing to jet power, and for the first time, middle-class passengers began regularly using airlines for leisure. As aviation advanced, it changed many areas, from politics to food. Before aviation, national leaders could not travel far abroad, but after planes became available global conferences were possible. Also, for the first time, fresh produce could be shipped across the world without rotting.

Atomic power was another technology from World War II, as the United States created atomic bombs to destroy the cities of Hiroshima and Nagasaki, forcing Japan's surrender. The first nuclear-powered submarine, the *Nautilus*, was launched by the US Navy in 1950. Beyond submarines and aircraft carriers, nuclear power plants generated electricity for entire populations, further transforming city life by yielding energy that did not come from coal-burning smokestacks. Nuclear power eventually began to be used for high-speed travel as well, such as the French TGV railway.

During World War II, US soldiers had come across lightweight European bicycles that were much more desirable than bulky American ones, which could weigh up to 60 pounds (27 kg), so US manufacturers started selling new, lightweight models. In the 1960s, American teens became obsessed with the Schwinn Stingray bicycle, with its long handlebars and small wheels. Ten-speed bikes hit the

The Evolution of Transportation Technology

The USS *Nautilus*, the first nuclear-powered submarine, enters New York Harbor. Nuclear submarines stayed underwater longer and cruised at higher speeds than diesel or electric submarines.

market as well, and between 1972 and 1974, bicycle sales doubled to fourteen million. Meanwhile, motorcycle riding was becoming more popular, as companies like Harley-Davidson branded themselves as part of a carefree or even "rebel" lifestyle.

THE RACE TO SPACE

The Cold War between the Soviet Union and the United States was an impetus for space travel. At the end of World War II, both countries brought in German scientists, who had built jet-powered weapons for their homeland. The US and Soviet militaries both desired intercontinental ballistic missiles based on the deadly German V-2 rocket. Their efforts eventually produced missiles carrying nuclear weapons capable of destroying life on Earth many times over. But fortunately, each superpower turned some of their military resources toward space exploration as well.

　　The Soviet Union launched its first satellite, *Sputnik I*, into orbit around the Earth on October 4, 1957. Not wanting to be outdone, the US government put together a space program that saw many successes and some dramatic tragedies as well. In an event that was televised globally, NASA launched Neil Armstrong and two other astronauts to the Moon on July 16, 1969. From the 1970s through the 1990s, the US program repeatedly launched space shuttles named after famous ships: *Columbia*, *Challenger*, *Discovery*, and *Atlantis*. Space travel eventually shifted away

Apollo 13, which returned to Earth after a nearly disastrous expolsion, took off from Kennedy Space Center on April 11, 1970.

from political struggles after the Cold War ended in the 1990s.

TRANSPORTATION ISSUES

Most twentieth-century versions of the car, plane, bus, tank, motorcycle, and helicopter were fuel-dependent, relying on fossil fuels. Cars made life easier for families in many ways but also caused huge traffic jams, pollution from exhaust, and accidents.

In the last few decades of the 1900s, changes were made to car production and safety that greatly improved health and well-being for drivers and passengers alike. Simple but effective shoulder-strap seatbelts, crash-absorbent "crumple zones" in the bodies of modern cars, and eventually, airbags became standard features. In the 2010s digital rear-view cameras and warning systems were common in new cars. In the twenty-first century, the challenge continued to be to develop transportation technology that was faster, environmentally sustainable, and accessible to a wider range of people.

Chapter 4

THE FUTURE OF TRANSIT

In the past, inventors were preoccupied with developing the first, fastest, or most powerful vehicles. Now communities, countries, companies, and individuals are shifting toward more sustainable ways of thinking about technology. With concerns about global climate change and awareness of the unequal treatment of different people, transit is evolving. New technologies related to computers, the internet, and artificial intelligence have opened new ways to move into the future.

MAKING TRANSPORTATION ACCESSIBLE

Until the 1990s in the United States, individuals with disabilities were not often accommodated in public spaces. Wheelchairs could not easily enter buses or

The Future of Transit

buildings with staircases at the entrance. People with visual or hearing impairments did not have information they could access readily on public transit.

In 1990, the Americans with Disabilities Act required that individuals with physical and mental disabilities be granted more equal opportunities. Buses were equipped with wheelchair lifts, and subway stations were required to have elevators, accessible bathrooms, and better audio-visual displays. Handicapped parking spots were designated for people with mobilty issues. Despite this progress, however, rural and suburban areas do not often have public transit options, and people with disabilities still encounter barriers for many reasons.

Public transportation has been adapted to fit the needs of people with mobility issues, providing more space, designated seating, and lifts for wheelchairs.

GREENER TRANSIT

As the general consensus among scientists has alerted us to global climate change, there is heightened concern about the toll that transit takes on the environment. Fuel-dependent vehicles deplete Earth's supply of petroleum, and they also emit large amounts of carbon dioxide and other greenhouse gases into the atmosphere. Global warming is primarily caused by the release of such gases, which trap heat in the atmosphere and warm the planet.

In response, countries have banded together at global councils to come up with ways to reduce pollution. The Kyoto Protocol brought 150 countries together in 1997 to pledge to lower their emissions. The United States and Australia were the only industrialized countries that failed to join the agreement, as their governments cited fears of financial penalties if emissions were not lowered. The Paris Agreement of 2015 brought together 195 countries that pledged to reduce greenhouse gases. Despite widespread criticism, the US president Donald J. Trump withdrew from the accord in 2017. However, many local and state leaders in the United States and elsewhere pledged support for these initiatives.

The Future of Transit

Meanwhile, in response to environmental concerns, electric cars and fuel-efficient "hybrid" cars are continuing to be developed. Tesla manufactures electric cars that can travel hundreds of miles before needing to be recharged. Hybrid cars, such as the Toyota Prius, combine gas and battery power so the battery can be recharged while braking—with energy that would have been lost as heat—and at various other times while driving. Smart cars, meanwhile,

The Toyota Prius is a commercially successful hybrid electric car, notable for its fuel efficiency and affordability.

are miniature cars recognizable for their petite size. Companies such as Google and Uber have begun testing self-driving cars, which could become even safer than ordinary automobiles in the near future.

An even greener option is the bicycle, and many cities have added bike lanes and bike protections to encourage safer urban biking. Minneapolis has a "bike freeway" allowing cyclists to traverse the city on specific bike paths. Many European cities have led the way in this regard. For example, Amsterdam also has a thriving bike culture, and biking is a way of life for many Dutch citizens.

ELECTRIC CARS AT TESLA

Tesla produced the first electric sports car in 2008 and a luxury sedan in 2012. The Model 3, released in 2017, was the first Tesla car designed to be affordable to the mass market at $35,000. Tesla cars can be powered at locations across North America, Europe, and Asia. They use thousands of small lithium-ion batteries instead of the large, single-purpose cells that most cars rely on. A Tesla factory in Nevada, called Gigafactory, produces a large proportion of the car batteries.

The Future of Transit

Tesla also designs solar panels and solar roofs, and it has pushed other corporations to follow its lead advancing renewable energy. Tesla's success is dependent on consumers' awareness of fuel dependency and global climate change, and its chief executive officer, Elon Musk, has attempted to promote scientific literacy and environmental stewardship among government and business leaders.

Tesla manufactures plug-in electric cars as well as other environmentally friendly technology, including solar panels.

TWENTY-FIRST CENTURY SPACE TRAVEL

Peaceful international cooperation appears to be the expectation for space exploration in the future. In the 1980s and 1990s, NASA was able to launch several satellites into orbit for communications companies. In addition, the Global Positioning System, or GPS, which relies on a network of Earth-orbiting satellites, was once restricted to military use, but it was opened up to civilians in the 1990s. As a result, GPS is now vital to air travel, smartphones, and car navigation systems throughout the world.

In 1998, Japan, the European Space Agency, Russia, Canada, and the United States's NASA program all joined together to build the International Space Station. By the early 2000s, civilian corporations

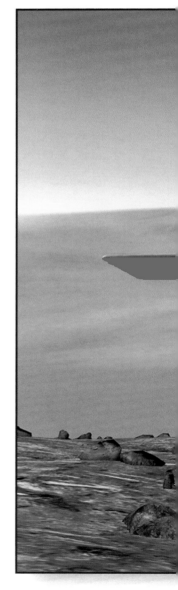

The Future of Transit

were also able to launch an increasing number of satellites into Earth orbit, in addition to those being launched by government and international agencies.

Mars rovers have been sent to Mars to collect data on the planet's livability and have found evidence of water there.

The Evolution of Transportation Technology

NASA and other space agencies have been particularly interested in exploring Mars to see if it could support life. NASA has sent several probes, including the Mars rovers *Curiosity*, *Spirit*, and *Opportunity*, to gather data about the planet's terrain. They found evidence of frozen subsurface water and alkaline soil, indicating that Mars might have been hospitable to life in the past. Meanwhile, as private companies have become more active in space exploration, Elon Musk of Tesla designed a private company, SpaceX, to make affordable rockets. SpaceX created the *Dragon* spacecraft, which helps deliver supplies to the International Space Station.

DRONES, SKATEBOARDS, AND HOVERBOARDS

Many forms of transportation are used for recreation, though some were not created for peaceful purposes. Drones are Unmanned Aerial Vehicles (UAVs) that developed out of military technologies. Most military drones have been piloted remotely, sometimes through computers thousands of miles away. Many now have autopilot capabilities, and experiments with drones and artificial intelligence have been conducted.

Drones have also been adapted for personal use, and small, remote-controlled drones can be used to take photos and videos from above. Most hover via a series of rotary, helicopter-like blades. Tourism agencies have used drones to film cultural wonders such as pyramids and mountaintop ruins. Scientists have used drones to monitor endangered species, and in 2017 drones dropped food pellets to save prairie dogs from starvation in Montana. Companies such as Amazon have begun to experiment with using drones to deliver goods to individual homes.

Skateboarding has been a sport and part of youth subculture since the 1960s, and people of various ages use skateboards for transportation on college campuses and city streets. Hoverboards are boards with wheels run by electricity. A person can stand on a hoverboard to be taken down the street. It does not require the rider to push with physical strength, unlike a skateboard or a scooter. However, some hoverboards have been recalled because of faulty designs that cause them to catch fire.

RIDE, CAR, AND BIKE SHARING

Access to transportation is also changing with the development of ride-sharing and vehicle-sharing programs. In the early twenty-first century the US corporation Uber revolutionized the taxi industry by allowing drivers and riders to connect via an internet app. Soon competitors such as Lyft and Juno also emerged with similar apps. Various start-up companies have also allowed people to rent cars in a cheaper and easier way. These ride-sharing and car-sharing programs have cut down the number of people who need to own or rent cars through large car rental companies.

Bike-sharing programs such as Citi Bike allow people to rent bikes from stations in urban areas and then return them to different stations. Most of these systems require an online reservation or a checkout process using a credit card or smartphone. By 2018 smartphones were being used for even more flexible, yet still experimental, bike-sharing systems, which allow a rider to find and/or leave a bike on a sidewalk, rather than having to walk to a particular bike station first.

Bike-sharing programs allow city dwellers to easily rent and secure bicycles, providing an alternative to cars, buses, and taxis.

TRANSIT CHALLENGES AND THE FUTURE

Innovation and access to technology is not evenly distributed across the world; in fact, many countries still do not have reliable public transit, and only their wealthier citizens have access to cars. A Pew Research survey of forty-four countries found that in parts of Africa and Asia, few people own cars, though they do own other types of vehicles. In Uganda, for instance, only three percent of people own cars, but 49 percent own bikes and 13 percent own motorcycles. And in Vietnam, two percent of people own cars, but 86 percent own motorcycles and 67 percent own bikes. In densely populated cities like Mumbai, India, people share the busy, congested streets on a mixture of bikes, scooters, livestock, cars, and buses. Traffic jams are almost legendary in Mumbai, however, so that people who do own cars might not benefit greatly from them.

In Nairobi, Kenya, public transit is dominated by *matatus*, individual buses that can take hours to reach destinations because of street congestion. Start-up culture is making its way to Nairobi, nonetheless, where *matatus* are being outfitted with Wi-fi and passengers can access up-to-the-minute traffic information so they can decide how they get to work.

In recent decades we have been moving faster and farther, and computers are now changing transportation at both local and global scales. The future of transit technology seems to be heading toward more green, inclusive, accessible, and pleasurable ways of getting around. Yet significant challenges will need to be overcome, including street congestion, overcrowding, and a lack of funding for public transit systems, if the world is to reap the full benefit of the transportation technologies being imagined today.

GLOSSARY

ACCESSIBILITY The degree to which a place can be reached or a vehicle can be operated by physically disabled or impaired operators.

APP A computer program or "application" usually designed for use on smartphones.

ASTROLABE An ancient hand-held instrument that uses the positions of celestial bodies in order to calculate the location of a ship.

AVIATION The flying, maintenance, and building of aircraft.

AXLE A pin or shaft on, or with which, a wheel or pair of wheels revolves.

DEFOREST To completely cut down or clear forests in an area.

EMISSIONS Substances discharged into the air, as by a smokestack or an automobile engine.

EXPLOIT To make use of something, such as a natural resource, for one's own advantage.

GREEN Relating to sustainable practices or environmentalism in everyday life or in political movements.

INFRASTRUCTURE The system of public works—such as highways, bridges, canals, power stations, electrical lines, and sewers—that helps develop a country or region.

Glossary

MAIZE The Native American term for corn, and now also the international word for corn.

MESOPOTAMIA Region of southwestern Asia between the Tigris and Euphrates rivers extending from the mountains of eastern Turkey to the Persian Gulf.

NATURAL RESOURCES Materials such as minerals and energy sources such as waterpower that are supplied directly by nature.

NUCLEAR ENERGY Energy that is created by splitting apart the nuclei of atoms, especially within electricity-generating nuclear power plants or a nuclear reactor that functions as an engine for a ship, submarine, or train.

PLANTATION An agricultural estate that usually includes housing for large numbers of laborers.

RECALL To summon the return of a commercial product because of safety concerns.

SATELLITE An object that orbits a star or planet, such as the Moon orbiting the Earth. An artificial satellite is an object people have launched into orbit.

SMARTPHONE A cellular phone that can run apps and connect to the internet.

SPOKE A small support bar radiating from the hub of a wheel to the rim.

The Evolution of Transportation Technology

START-UP A new business enterprise, such as a small company attempting to develop a series of new smartphone apps for commuters.

SUSTAINABLE Relating to a method of using a resource so that the resource is not depleted or permanently damaged.

FOR FURTHER READING

Bailey, Diane. *How the Automobile Changed History*. Minneapolis, MN: Essential Library, 2016.

Bellwood, Lucy. *Baggywrinkles: A Lubber's Guide to Life at Sea*. Portland, OR: Elea Press, 2016.

Green, Dan. *Eyewitness Energy*. New York, NY: DK Publishing, 2016.

Holl, Kristi. *Ancient Mesopotamian Technology*. New York, NY: Rosen Publishing, 2017.

James, Simon. *Ancient Rome*. London, UK: Dorling Kindersley, 2015.

Kirk, Ruth M. *STEAM Guides in Transportation*. North Mankato, MN: Rourke Educational Media, 2017.

Morgan, Ben, and Robert Dinwiddie. *Space!* New York, NY: DK Publishing, 2015.

Mulder, Michelle. *Pedal It! How Bicycles Are Changing the World*. Victoria, BC: Orca Book Publishers, 2013.

Nixon, Jonathan. *Energy Engineering and Powering the Future*. St. Catharines, ON: Crabtree Publishing, 2017.

Richardson, Gillian, and Kim Rosen. *Ten Routes That Crossed the World.* Toronto, ON: Annick Press, 2017.

Roy, Philip. *Eco Warrior.* Vancouver, BC: Ronsdale Press, 2015.

Wadhwa, Vivek, and Alex Salkever. *The Driver in the Driverless Car: How Our Technology Choices Will Create the Future*. Oakland, CA: Berrett-Koehler Publishers, 2017.

WEBSITES

European Union Transport Policy
https://europa.eu/european-union/topics/transport_en

International Space Station
https://www.nasa.gov/mission_pages/station/main/index.html
Facebook, Instagram: @ISS, Twitter: @Space_Station

United Nations Development Programme
http://www.undp.org/content/undp/en/home.html
Facebook, Instagram, Twitter: @UNDP

INDEX

A
Africa, 8–10, 15, 17–18, 20, 56
airplane/jet, 31, 33–39, 41, 43
airship, 35–36
animals, 4, 6, 8–9, 11–13, 20–22, 24, 53
Armstrong, Neil, 41
artificial intelligence, 44, 53
Asia, 8–9, 15, 17, 19–20, 48, 56
atomic/nuclear power, 39, 41
automobile/car, 4, 30–32, 36–38, 43, 47–48, 50, 54, 56
   early models, 31–34
   highway system, 38
   hybrid, and smart car, 47

B
bicycle, 48, 54, 56
   early inventions, 25, 27–28, 34
   lightweight models, 39–40
boat/ship, 12–18, 20–22, 24, 28–31, 41

C
city/public transit, 4–5, 24–25, 44–45, 56–57
Cold War, 41, 43
Columbus, Christopher, 15

D
deforestation, 16–17
drones, 52–53

E
Earhart, Amelia, 34–35
Egypt, 8–10, 13
Europe, 10–11, 14, 15–18, 20, 22, 25, 27, 29–31, 34, 38–39, 48, 50

F
Ford, Henry, 33
Fulton, Stephen, 21–22

G
Germany, 32–33, 37–38, 41
global climate change, 44, 46, 49
Global Positioning System (GPS), 50
Great Britain, 16, 19, 21–22

H
highway systems, 4, 11, 38
Hindenburg, 36

63

The Evolution of Transportation Technology

I
indigenous peoples, 11–13, 17–18
Industrial Revolution, 19–22, 24–25, 27–29
International Space Station, 50–52

J
Japan, 37, 39, 50–51
Jeep, 37

L
Leonardo da Vinci, 33
Lindbergh, Charles, 34

M
Middle East, 6, 8–11
motorcycle, 40, 43, 56
Musk, Elon, 49, 52

N
NASA, 41–43, 50, 52
Native Americans, 11–13
Nautilus, 39

O
Olds, Ransom E., 33

R
railroad, 20, 22–24, 28–29
rocket, 4, 31, 41, 52
Roman Empire, 10–11
Russia (Soviet Union), 41, 50–51

S
South America, 8–9, 11, 20, 29
space exploration, 30, 41, 42, 50–52
Spirit of St. Louis, 34
Sputnik I, 41
Stanley, Francis and Freelan O., 31–32
steam power, 4, 20–22, 24, 28–29, 31–32
steamship, 21–22, 28–29
submarine, 4, 39
subway, 24–25, 45

T
tank, 37–38, 43
Tesla, 47–49, 52
truck, 31, 38

W
wheel, 8–11, 20, 25, 39, 53
World War I, 31, 34–35, 38
World War II, 31, 37–39, 41
Wright Brothers, 33–34